AQA A2 Biology:

Writing the Synoptic Essay

by

Dr Robert Mitchell

CT Publications

A catalogue record for this book is available from the British Library

ISBN 978-1-907769-01-6

First published in May 2010 by
CT Publications

Published in 2010 by
*CT Publications**
40 Higher Bridge Street
Bolton
Greater Manchester
Bl1 2HA

Edition 10 9 8 7 6 5 4 3 2 1

*CT Publications is owned by *Chemistry Tutorials* located at the same address.

CONTENTS

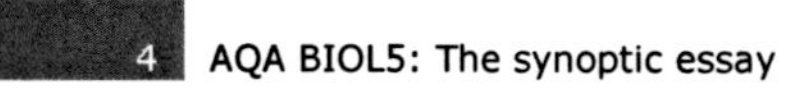

1. BE AWARE OF THE TASK YOU FACE

- ✓ You are expected to construct an essay of between 700 and 1100 words covering a diverse range of synoptic content from the entire A-level specification.
- ✓ Your essay is assessed on a scale (see Appendix 1) marked for content (16 points), breadth (3 points), relevance (3 points) and quality of language (3 points)
- ✓ You will have to select an essay from a choice of 2 titles

WHAT IS SYNOPTIC?

Syn-op-tic [Si-nop-tik] Pertaining or constituting a synopsis; affording or taking a general view of the principal parts of a subject.

This definition comes from *www.dictionary.com* and gives a reasonably good sense of the meaning. Essentially at A-level it can be assumed to mean a focus on key underlying concepts in the subject that are reflected throughout at all levels. So for example in biology, all organisms use ATP to release energy for processes and use proteins as enzymes to control specific reactions. In this context these concepts are deemed to be synoptic and will be covered frequently in questions.

WILL I HAVE TO LEARN EVERYTHING?

Simply put ... ***NO!*** Synoptic questions and essays are not about the regurgitation of every last fine piece of detail, but focus on a DRAWING TOGETHER *OF KEY SYNOPTIC ELEMENTS*. In this book I have identified these elements and provide a process you can easily follow to internalise and consolidate them.

A SYNOPTIC CARBON ATOM

*As a tool to illustrate synoptic thinking I want you to consider a carbon atom, say floating in the air in a CO_2 molecule soaking up infra-red rays in the 1700 cm^{-1} region from the Earth and thereby doing its bit for global warming. Let's follow the fate of this atom and see how many key synoptic elements we can identify (these are highlighted in **bold** for you).*

The infra-red rays strike the **molecule** giving it **kinetic energy** to excite it and make it **move** more rapidly. This allows it to come into close proximity to a leaf of a dandelion plant. Here it **diffuses** through the open stomata down a **concentration gradient** into the air spaces between the mesophyll **cells** and enters the palisade **cell** where it **diffuses** through the **plasma membrane** and through the cytoplasm and on into the stroma of an **organelle** called the chloroplast. Here it is combined to a 5-carbon **sugar**, RuBP, broken down to glycerate-3-phosphate and then gets **reduced** by **hydrogen ions** and **electrons** by a reduced NADP **coenzyme molecule**. Now as an atom in triose phosphate, the **carbon** atom then undergoes a **condensation reaction** to form **glucose**, a **monosaccharide**. The **glucose** it then joined to a fructose to form the

disaccharide sucrose which is **actively transported** into the phloem vessels though a **specific channel protein** using **energy** from the **hydrolysis** of **ATP** to provide **energy** to push it against a concentration **gradient**. This lowers **water potential** allowing **water** to enter the phloem by **osmosis** creating a **hydrostatic pressure** that forces the solution of sugar towards the roots of the dandelion. There it becomes **hydrolysed** to glucose by **enzymes** and **condensed** into a cellulose **polysaccharide** molecule and **assimilated** into the **cell** wall of a root hair **cell.** In doing so, the **carbon** atom is incorporated as part of the plants **biomass** and forms a small part of the **productivity** for the **ecosystem**.

Presently, the dandelion's roots are **consumed** by a **primary consumer** such as a sheep. The sheep swallows the cellulose and the **saprobiotic microorganisms** in its digestive **system** release **enzymes** which **catalyse** the **hydrolysis** of the cellulose into β-glucose. The glucose is **absorbed** through a sodium-glucose co-transporter **protein** through a **specific channel protein** against a concentration **gradient** where it passes into the sheep's **blood**, lowering its **water potential** prompting the release of the **protein hormone** insulin. The insulin and the synoptic carbon atom inside the glucose **molecule** are then **transported** though the blood inside the cardiovascular **system**, down a **pressure gradient** caused by the **contraction** of the ventricles in the sheep's heart. Upon entering a capillary leading into an **organ**, in this case, the liver, the **hydrostatic pressure** forces the glucose, insulin and other small molecules into the **tissues** through the formation of a **tissue** fluid. The insulin **hormone binds** to a **specific complementary receptor** on the **plasma membrane** which opens a **specific** glucose **channel** allowing the glucose to enter the hepatocyte by **facilitated diffusion** across the hydrophobic phospholipid barrier.

Inside the hepatocyte, the glucose becomes **phosphorylated** by **ATP** and undergoes **glycolysis**, being **oxidised** by dehydrogenase **enzymes** to pyruvate. This **diffuses** into the matrix of a mitochondrion, an **organelle,** where it undergoes decarboxylation forming CO_2 (with our synoptic carbon atom) which **diffuses** out though the **lipid bilayer** and ultimately out of the **cell** and through the endothelial cells into the lumen of a capillary. Upon **dissolving** in the **water** of the blood plasma producing HCO_3^- **ions** and, it **diffuses** into a red blood **cell** where it binds to a Fe^{2+} **ion** on a haemoglobin **protein** molecule, unloading **oxygen** in the process at the **exchange surface**. The red blood cell is then **transported** by the beating of the heart to the lungs where the CO_2 **molecule** diffuses off the haemoglobin (as **oxygen** loads in its place) and through the one cell thick epithelium **tissue** of the alveoli into the air in the lungs. **Relaxation** of the intercostal **muscle's** actin and myosin **protein** filaments and the arching of the diaphragm **move** the rib cage down and in, decreasing the volume of the thorax and increasing the **pressure** in the lungs to higher than atmospheric. This creates a pressure **gradient** that forces the CO_2 molecule back though the respiratory **system** and back out into the atmosphere.

Now consider this; we have just written a short synoptic essay! It was about seven hundred or so words covering many synoptic elements. This essay could have been a response to the title ...

Describe the processes by which a carbon atom is transferred between organisms, and between organisms and their environment.

The content covered at least three kingdoms (plants, animals and microorganisms), drawing on content from all four modules (e.g. osmosis and carbohydrate and digestion, module 1), photosynthesis and respiration (module 4) and oxygen dissociation and gas exchange (module 2). In the following sections I will describe the precise nature of the synoptic elements covered, and others, and show you a process to follow in order to start to learn them.

WHICH CONTENT IS SYNOPTIC?

Biomolecules

- ✓ A list of biological molecules containing different chemical elements
- ✓ Structure and importance of specific molecules
- ✓ Roles and specific examples of specific molecules

Enzymes

- ✓ Activation energy
- ✓ Generic examples
- ✓ Plant specific
- ✓ Animal specific

Genetic code

- ✓ Structure of DNA
- ✓ Nature of the genetic code
- ✓ Replication
- ✓ Transcription
- ✓ Translation
- ✓ Mutations
- ✓ Roles in differentiation
- ✓ Roles in intra- and inter-specific variation
- ✓ Transfers of genes (horizontal/vertical/sexual)

Cells

- ✓ Plants
- ✓ Animals

Microbes

- ✓ Beneficial roles
- ✓ Harmful roles

Gas exchange

- ✓ Features of gas exchange
- ✓ Surface area to volume ratio good for gas bad for H_2O loss

Movements

- ✓ Diffusion
- ✓ Osmosis
- ✓ Active transport
- ✓ Endocytosis
- ✓ Exocytosis
- ✓ Phagocytosis
- ✓ Hydrostatic pressure
- ✓ Muscle contraction

Energy

- ✓ ATP
- ✓ Heat
- ✓ Light
- ✓ Chemical
- ✓ Photosynthesis
- ✓ Aerobic respiration
- ✓ Anaerobic respiration

Adaptations

- ✓ Xerophytes
- ✓ Behavioural
- ✓ Physiological n(haemoglobin etc)
- ✓ Anatomical
- ✓ Gas exchange and cell shapes
- ✓ Natural selection
- ✓ Speciation

Diversity

- ✓ Interspecific and intraspecific variations
- ✓ Continuous/discontinuous
- ✓ Diversity index
- ✓ Stabilisation of food webs
- ✓ Survival

Coordination/communication

- ✓ Hormonal
- ✓ Nervous
- ✓ Courtship
- ✓ *taxes*

HOW TO USE THE16:3:3:3 RULE

The examiners are looking for the following breakdown:

- ✓ 16 Content marks drawn from either specification content or a-level equivalent knowledge.
- ✓ 3 Breadth marks based on either 3 different modules content or 3 different kingdoms of examples.
- ✓ 3 Relevance marks – assume you have them in the bag but lose one each time you make an error.
- ✓ 3 Quality of communication – learn how to use capital letters, full stops and commas appropriately.

CONTENT

The content you include should span at least 5 areas in some detail with two key points and a specific example. So if I was being asked for an importance of ATP, I would state that ... *its **hydrolysis by ATPase releases energy** that can be used for **specific** processes, for example **provides energy** for the **active transport** of **mineral ions against a concentration gradient** in a **root hair cell** of a **plant**.*

Notice just how many key terms there were in that sentence. These identify it as being high quality for the marker. I could have said ... *ATP is used for active transport of ions in a plant ...* technically the same point but with a much lower quality.

BREADTH

These should be the easiest to gain, just make sure you use examples from at least 3 modules or kingdoms. Remember that most of the examples are animal, fewer plant and hardly any fungi and bacteria ... and most of my students struggle to identify the other kingdom, let alone give examples from it.

But you can collect **generic information** ... by that I mean processes that are common to all living organisms. For example all organisms have a phospholipid bilayer for a membrane that is embedded with channel proteins. So if you were doing an essay on proteins and had very few bacteria examples you could source the membrane channels as an example and illustrate with a bacterial example. E.g. **Protein** molecules can act as **channel** proteins which provide a route by which **hydrophilic** species can pass **through** the **cell membrane**. A **bacteria** such **E. Col**i in the **stomach** of a **mammal** would use these channels to **absorb nutrition** such as **ions** like **Na^+**, from the **host**.

RELEVANCE

Essentially the only advice here is to keep to the point. In an essay on role of nucleic acids for example, talking about the job of the proteins they code for is off-topic and will **LOSE** you marks. Also common errors like saying DNA is made from polypeptide strands or that DNA is a protein will lose you these marks. **STICK TO THE POINT**.

QUALITY

I should not have to say much here, but quality goes beyond physical sentence construction. Think about the message you are conveying. Lead the reader through a logical sequence. Essays that flit about pointlessly and are incoherent are a nightmare to mark. Plants one minute ... animals ...plants ....fungi...animals...plant...animal...damn it makes my ears bleed to mark them.

Get an overview of what you are going to write... and use it as a template. Remember the PPPPPP rule ... Planning and Preparation Prevents Piss-Poor Performance!

2. SCANNING A TITLE FOR CLUES FOR EFFECTIVE PLANNING

- ✓ Use a **head NOT heart** approach. Do not be drawn into a title you ***like*** if it not broad enough to gain easy marks. Many essays quickly dry up. For example "**negative feedback in living organisms**" is actually quite limited, even if you feel you can do a reasonable job from that modules content.
- ✓ Essays of the 50-50 type are much more open as they allow a two-pronged attack of the title (see below).
- ✓ Look for any words that allow an expansion of content. For example "**living organisms**" allows you to draw on animals, plants, bacteria, fungi or protoctist examples.

3. PLANNING TO GAIN MARKS

Plan according to the title, carve it up into about 8 by 2 units, or 4 by 4, and then get started!

Previously I showed you broadly how to aim to gain marks with the 16:3:3:3 rule, and now we'll look at the planning stage. Take a look at any synoptic essay title and you'll notice that they fall into two main categories.

THE 50-50 SPLIT, EXAMPLES OF THESE INCLUDE:

- ✓ The structure **and** functions of carbohydrates.
- ✓ The causes of variation **and** its biological importance.
- ✓ The rising temperatures may result in physiological **and** ecological effects on living organisms.
- ✓ Cells are easy to distinguish by their shape. How are the shapes of cells related **to their** function?

See how each essay can be carved up into a simple split – *something* ***and*** *something* or a *relation* ***to its/their*** *whatever...* etc

Let's try to carve up the first title, the **Structure and functions of carbohydrates**. On a piece of paper, split the page in two vertically and at the top on the left put structure and on the right, function. Now split the page into eighths so your paper is a grid of **16 boxes**. NOTICE its **16 boxes!** And earlier we saw that we needed **16 content points**. Now in your plan you could fill the boxes in the functions and relate structure to them. For example:

- ✓ [on right] cellulose cell walls provide strength and support & [on left] long unbranched/polysaccharide/hydrogen bonds/microfibrils
- ✓ [on right] component of DNA/RNA nucleotides & [on left] monosaccharide/5-Carbon sugar distinguishes DNA from RNA.
- ✓ [on right] energy storage molecule in plants (starch) & [on left] branched, large amounts of glucose in small volume/easily hydroysed/insoluble.

This process can be repeated until as many of the boxes as possible are filled. Once you have collected something to say on each of the 8 pairs, you can structure and begin the essay.

THE SHORT AND SWEET TYPE, EXAMPLES ARE

- ✓ Movements **inside** cells
- ✓ Negative feedback **in** living organisms.
- ✓ Transfers **through** ecosystems.
- ✓ Cycles **in** biology.

Notice how these can be recognised by *a process* ***in/through/between*** something. At first glance these seem quite daunting, too wide (or too narrow depending how unprepared you are!) I like these as they provide a wide canvas to prepare a plan on.

Let's take say **"Cycles in Biology"** Our biology has been studied at a wide number of levels – molecular – cellular – organs – systems – organisms – populations – communities – biotic – abiotic – environmental. Notice there are nine levels here, so aiming for 2 cycles in each would fill our **8×2=16** grid nicely and allow us to start the essay. So, I would move along each level and find a cyclical change. Some levels I may struggle to find any so I would miss it and try to find an extra one elsewhere.

- Molecular – [ATP to ADP and Pi and back] [NAD + $2H^+$ + $2e^-$ forming reduced NAD and back]
- Cellular – [The cell cycle] [Krebs] [Calvin]
- Systems -[breathing cycle] [cardiac cycle]
- Organisms – [life cycle] [menstrual cycle]
- Populations – [lag-log-stationary-death-repeat]
- Communities – [predator prey]
- Abiotic -[night-day] [Seasons]
- Environment [Nitrogen] [Carbon cycle] [Energy]

Notice how the simple act of dividing up the essay has broken it down a much easily digestible task and we automatically build in breadth!

The key to the above types of essay is to find that hook, that point that allows you to carve up the roast so as to lever the topic open to gain the breath. There are trigger words in the titles that allow this

- ✓ **Living organisms** – find examples from animal – plant – fungus – bacteria – protoctist – generic
- ✓ **Biology** – use the levels I used in the example above
- ✓ **Cells** – use examples from eukaryotic plant and animals as well as prokaryotic
- ✓ **Ecosystems** – think abiotic, biotic and trophic levels, also material and energy transfers

So in summary I have shown you how to quickly make a broad plan, an outline for the framework of the essay that uses the title itself to pries open the lid to reveal contents – the meat of the essay.

4. ORGANISING THE PERFECT ESSAY

In the last section I described how to carve up the essay using clues within the title so it became easier to handle and more defined. Now I want to look at developing that theme so that we can add detail and produce an essay that flows and has structure instead of a random series of unconnected sentences.

STRUCTURE

If you have done what I suggested then you will have a piece of paper on which you plan is outlined, but it will be carved up into a series of boxes. It may be that you were unable to find examples to enter for some boxes – that's ok because often more things occur as we proceed with the development of the essay. Now take an objective look at the grid and assess the following:

IS IT BALANCED?

1. By that I mean are there whole sections with nothing in them – if this is the case you must try to think of content to fill them.
2. Try not to keep adding more examples into boxes – your focus MUST be on putting something ... anything in as many boxes as possible. Once you have made a point, you will not get any more by using more examples. Common errors include giving about 10 examples of enzymes on a protein importance essay – enzymes are after all, only ONE role of a protein.
3. If you struggle to find specific content, then try to use generic things from biology, like cells, membranes, ATP, proteins, enzymes, respiration, competition, need for nutrition, variation, etc to **force** specific examples. E.g. if I needed a role of a protein in a fungi I can always use a generic example of, say the enzyme ATPase which will hydrolyse ATP to ADP and Pi releasing energy for active transport for mineral ions into a hyphae of a saprobiotic fungus.

IS THERE A NATURAL ORGANISATION?

- ✓ Essays often split into broad **sub-categories** like Animal, Plants etc or Physiological and Ecological effects. Try to find sub-categories to give your essay flow – imagine it a little like blocks stuck together – the **animal** block then the **plant** block then the **other organism** block.
- ✓ Once you identify the sub-categories and have enough boxes filled I suggest you finalise the content and start to write.
- ✓ Use the information from one of your boxes at a time and try to develop it into the following elements. Try to expand the simple statements in the boxes like **NAD** into ... **2 key points** ... at least **2 specific examples** from the specification where appropriate. So if I was doing Cycles in Biology, and I was in the molecular box, I might say something like *" ... co-enzymes are molecules that work with dehydrogenase enzymes [key point 1]to transfer hydrogen ions and electrons from a substrate during respiration [key point 2] for example NAD accepts $2e^-$ and $2H^+$ from glucose during glycolysis in an animal cell during aerobic*

respiration [specific example 1] and NADP accepts $2e^-$ *and* $2H^+$ *during the light dependent reaction in the thylakoid of a chloroplast in a palisade cell during photosynthesis [specific example 2]. Both of these get re-oxidised when the electrons are passed down an electron transport chain [the cycle bit]"*

✓ Always, always, always link the content back to the title of the essay. Avoid falling into the trap of just regurgitating lists of flow charts of content. So if I was doing an essay on "*DNA and the transmission of information*" I would not just discuss the key points of DNA replication or mitosis, I would link them back to the title first to show the examiner I understood the context. For example, I would put them in a context like ... "*The genetic information encoded in DNA must be transmitted to genetically identical daughter cells when mitosis takes place for growth or repair. To ensure this takes place, the semi-conservative replication of DNA takes place in late interphase of the cell cycle. The enzyme helicase ...*" and then I could include the detail of replication and mitosis safe in the knowledge the context had been made clear.

5. WHAT TO DO IF YOU'RE STUCK

This is a nightmare, you've gotten half way through and you lose the plot a little and dry up! What can you do? Here are a few tips to re-open the essay:

✓ Force yourself to think outside the box, try to come up with any relevant thread you can.
✓ Add extra if you dry up by asking yourself ... ***WHY***? or ***SO WHAT***? or ***TO DO WHAT***? These are great openers!
✓ Go back and add detail, a few words here and there can increase content marks.
✓ Avoid sweeping statements like ... fur is important to animals ... say precisely WHY it is IMPORTANT (if offers a means of camouflage and protection from predators for example) and ask yourself what is the CONSEQUENCE of NOT having it?
✓ Avoid going in to deep – keep to *BREADTH NOT DEPTH*.
✓ Back up every statement with an appropriate example from the syllabus ... don't be shy of using a GCSE example, as its better than nothing.
✓ Avoid lengthy openers or conclusions – these essays are NOT English essays, every word is there for one reason ... to gain marks.
✓

6. TIMING

If you have **16 boxes** and about 35 minutes of writing with 10 minutes planning time it follows you spend about **2 minutes per box**. Remember this – **BREADTH NOT DEPTH** in these essays, go for lots of examples that skate the surface of your biology in a connected way – this is that they are looking for – not loads of fine detail about one or two things!

FLESHING OUT THE PLAN IN THE LAST FEW MINUTES

It may seem odd to discuss adding flesh to the bones of the plan at the end, but it is **VITAL**. The examiners **WILL CREDIT** content that is clearly described in the plan, but only if it has not been included. This is why we did not spend ages planning. Now ... **IN**

THE LAST 5 MINUTES ... we go back and add as much clarity and detail to the plan as possible, as these extra marks could be the make or break of our essay.

7. CHOOSING THE RIGHT ESSAY FOR YOU

Let's be clear about what you will face. The essay comes off the back of a big paper and about 10 marks of data handling. You will then face the "open" titles that I've included here and you will have about 45 minutes to write your opus. That's not easy. So now you are faced with a choice of TWO titles. Pick the wrong one and there is not enough time to go back and correct your mistake. So I want to focus on how to choose the right essay for you – and **choose the one with the greatest mark potential.** That should be your **only** criteria. It is important that you are not swayed by personal preferences. Imagine you loved horses and an essay came up ... Describe the role of horses in an ecosystem ... you'd be over the moon, off you'd go ... but really, where to? After you covered eating grass and excreting over a field, I suspect you'd run out of ideas!

I suggest you take an objective look at both titles and try to try to a speed plan, a quick carve up of both essay titles. Which has more scope..? Which one can you fill in more boxes for...? Which one has more detail that you know..? Look out for the following:

CLUES IN THE TITLE

Looking at the list of essay titles I included in Appendix 2, the following is a list of triggers in the titles which suggest it can be "opened up" to flood the marks.

- ✓ "Biology" or" Biological": This gives you opportunity to draw from a wide range of levels and/or kingdoms
- ✓ The 50-50 splits offer an easy entry into the carving-up process to start to digest a bigger essay into bite-sized pieces
- ✓ "Living Organisms" offers an easy splitting into animals, plants, fungi, bacteria and protoctists and/or generic examples
- ✓ "Transfer" or "Flow". Think of a movement or flow ... FROM...INTO...THROUGH...OUT OF...BETWEEN – again, a nice split to get you thinking.

TRIGGER WORDS TO BE WARY OF IN MY OPINION

- ✓ **"Importance"**. This is hard to get for most people, and I have a simple suggestion for deal with essays with *importance* in the title ... **avoid them!**
- ✓ "Negative Feedback" or "Osmosis" any **single narrow topic** – this could soon cause you to **dry up and run out of content** unless you are experienced at forcing examples.
- ✓ "Relationships" e.g. a Structure-fuction *relationship* of protein etc ... are do-able but you must take care as you must constantly connect and link the two parts otherwise the content and relevance marks could be very low.

8. PUTTING IT ALL TOGETHER

By definition, I suppose a good essay is one which scores 25/25 marks! If you have followed the tips outlined in the previous section you should be well on the way to generating one. There are some common features of essays that stand out in the examiners eyes and a prior knowledge of these allows them to be built into your essay. In general, a good essay will ...

- ✓ **Start with a definition of a title keyword.** Definitions show precision and knowledge and demonstrate a good understanding of the content. Starting with one sets the context of the essay and give a solid entry into the material. For example, if our essay title was PROTEINS AND THEIR IMPORTANCE TO LIVING ORGANISMS, a good definition opener could be ... *Proteins are polymers of amino acids joined by peptide bonds which carry out have a wide variety of key functions in living organisms...*
- ✓ **Follow up with a statement of intent.** A statement of intent demonstrates clear a focus for the essay and details the objective in a clear and concise manner. Continuing the example above we could now add ... *This essay will list and detail many of these specific roles in plants, animals, fungi and bacteria and highlight the benefits these molecules confer to the organism.*
- ✓ **Have a clearly discernible structure.** The precise structure will follow from the planning stage and be specific to different essays. Using the example above, it would be logical to follow a sequence of functions through the different kingdoms in turn, separating out those functions that were specific to each, e.g. pancreatic amylase which hydrolyses starch to maltose in the small intestine of an animal would not be mixed with in the same section with penicillinase, an enzyme released by a fungus that hydrolyses cell walls of bacteria.
- ✓ **Have breadth not depth**. One mistake many students make is to go into a topic in way too much detail. They fail to accept the point at which they scored the marks that were available for the points they are making. When studying the mark schemes of previous essays it becomes clear that only a few key points are required to score marks, but that many topic areas need to be covered. Using the example above, a good essay would cover a few specific examples of many roles for proteins like enzymes, hormones, receptors, fur, channels, antigens, antibodies etc rather than many examples of just hormones or enzymes.
- ✓ **Show an understanding of the title and not just repeat it**. In essays such as the one above, it is common to find phrases like "... receptors like the insulin receptor have a specific shape to recognise only insulin and so they are a really important role for proteins." This is telling the examiner it is important rather than demonstrating to him you understand why. Instead a phrase " ... the insulin receptor's complementary shape to insulin ensures a cells specific response to the hormone causing the glucose channel to open only when the concentration is high enough."
- ✓ **Uses specific content from the syllabus.** This should be an obvious point, but a clear demonstration of syllabus specific material will always impress the examiner and targets mark points you know will be there. E.g. The specification lists ... "*Enzymes as catalysts lowering activation energy through the formation of enzyme-substrate complexes.*" So in our essay we could stress that ... "*one crucial role for proteins is to act as enzymes. These are protein catalysts that control specific chemical reactions, increasing their rate by lowering the activation energy. This is*

brought about by the formation of an enzyme-substrate complex in which an active site binds to the complementary shaped substrate molecule."

- ✓ **Show evidence of extra depth of study**. Try to bring in some examples from the *How Science Works* aspects of the text book or link some content to your genera; knowledge of studies of other subjects. Statements like "*In a recent article in New Scientist it was suggested that ...* " will unconsciously help the examiner rate you as someone who has read around the biology and related subjects.
- ✓ **Uses definitions**. Following on from above, specific definitions of terms will always be a powerful addition to any essay. E.g., "*antibodies are Y-shaped globular proteins secreted by plasma B-cells, with a variable region that targets and binds specific antigens on a pathogen." ...* is probably more mark-worthy than "antibodies bind to and inactivate pathogens."
- ✓ **Contains specific named examples**. Always add quality and flesh to your essay with named examples or enzymes (maltase, DNA polymerase, helicase, etc), hormones (ADH, FSH, insulin etc), animals (seal, lions etc), plants (oak trees, dandelion etc), fungi (bread mould, yeast) .
- ✓ **Has clear flow and direction**. This type of essay avoids simply being a list. The reader is lead and guided in a planned sequence through the maze of content.

APPENDIX 1: AQA'S PUBLISHED GENERIC MARK SCHEME

General principles for marking essay questions

Four skill areas will be marked:

Scientific content (S)
Breadth of knowledge (B)
Relevance (R)
Quality of written communication (Q)

These skill areas are marked independently of each other. Providing that there is sufficient evidence, and the subject content is relevant to the question answered, it is possible for candidates to obtain maximum credit for skill areas B, R and Q, even if they gain little credit for Scientific content.

The following descriptors will form the basis for marking.

Scientific content (Maximum 16 marks)

Mark	Descriptor
16	Material accurate and of a high standard throughout, reflecting a sound understanding of the principles involved and a knowledge of factual detail fully in keeping with a programme of A-level study. In addition, there are some significant references to material which indicates greater depth or breadth of study.
14	
12	Most of the material is of a high standard reflecting a sound understanding of the principles involved and a knowledge of factual detail generally in keeping with a programme of A-level study. Material accurate and free from fundamental errors, but there may be minor errors which detract from the overall accuracy.
10	
8	A significant amount of the content is of appropriate depth. Shows a sound understanding of most of the principles involved and a knowledge of factual detail generally in keeping with a programme of A-level study. Most of the content is accurate with few fundamental errors.
6	
4	Material presented is largely superficial with only occasional content of appropriate depth. Shows some understanding of some of the basic principles involved. If a greater depth of knowledge is demonstrated, then there are many fundamental errors.
2	
0	Such material as is relevant is both superficial and inaccurate, rarely demonstrating evidence of knowledge in keeping with a programme of A-level study.

Note: Only 0, 2, 4 marks etc. are awarded. This limits the number of categories and improves the consistency of marking.

Breadth (Maximum 3 marks)

Mark	Descriptor
3	A balance account making reference to most areas that might realistically be covered in an A-level course of study.
2	A number of areas covered but a lack of balance. Some topics essential to an understanding at this level not covered.
1	Unbalanced account with all or almost all material based on a single aspect.
0	Material entirely irrelevant.

Relevance (Maximum 3 marks)

Mark	Descriptor
3	All material presented is clearly relevant to the title. Allowance should be made for judicious use of introductory material.
2	Material generally selected in support of title but some of the main content of the essay is only of marginal relevance.
1	Some attempt made to relate material to the title but considerable amounts are largely irrelevant.
0	Material entirely irrelevant or too limited in quantity to judge.

Quality of written communication (maximum 3 marks)

Mark	Descriptor
3	Material is presented in clear, scientific English. Technical terminology has been used effectively and accurately throughout.
2	Account is logical and generally presented in clear, scientific English. Technical terminology has generally been used effectively and is usually accurate.
1	The essay is poorly constructed. Often fails to use an appropriate scientific style and terminology to express ideas.
0	Material entirely irrelevant or too limited in quantity to judge.

Total 25

Examiners are looking for

- evidence of knowledge and understanding at a depth appropriate to A level
- selection of relevant knowledge and understanding from different areas of the
- specification
- coverage of the main concepts and principles that might be reasonably be expected in relation to the essay title
- connection of concepts, principles and other information from different areas in response to the essay title
- construction of an account that forms a coherent response
- clear and logical expression, using accurate specialist vocabulary appropriate to A level

Assessing Scientific Content

Maximum 16 marks.

Descriptors are divided into 3 categories: Good (16, 14, 12), Average (10, 8, 6) and Poor

(4, 2, 0). Only even scores can be awarded, i.e. not 15, 13, etc.

Examiners need first to decide into which category an essay comes.

A good essay

- includes a level of detail that could be expected from a comprehensive knowledge and understanding of relevant parts of the specification
- maintains appropriate depth and accuracy throughout
- avoids fundamental errors
- covers a majority of the main areas that might be expected from the essay title. (These areas are indicated in the mark scheme. Occasionally a candidate may tackle an essay in an original or unconventional way. Such essays may be biased in a particular way, but where a high level of understanding is shown a high mark may be justified.)
- demonstrates clearly the links between principles and concepts from different areas.
- Note that it is not expected that an essay must be 'perfect' or exceptionally long in order to gain maximum marks, bearing in mind the limitations on time and the pressure arising from exam conditions.

An average essay

- should include material that might be expected of grade C/D/E candidates
- is likely to have less detail and be more patchy in the depth to which areas are covered, and to omit several relevant areas
- is likely to include some errors and misunderstandings, but should have few fundamental errors
- is likely to include mainly more superficial and less explicit connections

A poor essay

- is largely below the standard expected of a grade E candidate
- shows limited knowledge and understanding of the topic
- is likely to cover only a limited number of relevant areas and may be relatively short
- is likely to provide superficial treatment of connections
- includes several errors, including some major ones
- Having decided on the basic category, examiners may award the median mark, or the ones above or below the median according to whether the candidate exceeds the requirements or does not quite meet them.

APPENDIX 2: AQA'S ESSAY TITLE LIST

Paper	Biology A	Biology B
Jun 02	The different ways in which organisms use ATP. How the structure of cells is related to their function.	The different ways in which organisms use ATP. How the structure of cells is related to their function.
Jan 03	How bacteria affect human lives. The biological importance of water.	The biological importance of water. The movement of substances within living organisms.
Jun 03	The structure and functions of carbohydrates. Cycles in biology.	The structure and functions of carbohydrates. Cycles in biology.
Jan 04	How carbon dioxide gets from a respiring cell to the lumen of an alveolus in the lungs. How an amino acid gets from protein in a person's food to becoming part of a human protein in that person.	How the structure of proteins is related to their functions. The causes of variation and its biological importance.
Jun 04	The transfer of energy between different organisms and between these organisms and their environment. Ways in which different species of organisms differ from each other.	The process of osmosis and its importance to living organisms. Energy transfers which take place inside living organisms.
Jun 05	Inorganic ions include those of sodium, phosphorus and hydrogen. Describe how these and other inorganic ions are used in living organisms. Bacteria affect the lives of humans and other organisms in many ways. Apart from causing disease, describe how bacteria may affect the lives of humans and other organisms.	Negative feedback in living organisms. Mean temperatures are rising in many parts of the world. The rising temperatures may result in physiological and ecological effects on living organisms. Describe and explain these effects.

Jun 06	Polymers have different structures. They also have different functions. Describe how the structures of different polymers are related to their functions. Describe how nitrogen-containing substances are taken into, and metabolised in, animals and plants.	The transfer of substances containing carbon between organisms and between organisms and the environment. Cells are easy to distinguish by their shape. How are the shapes of cells related to their function?
Jun 07	**Carbon dioxide in organisms and ecosystems.** **Why offspring produced by the same parents are different in appearance.**	**Movements inside cells.** **Transfers through ecosystems.**
Jun 08		The part played by the movement of substances across cell membranes in the functioning of different organs and organ systems. The part played by enzymes in the functioning of different cells, tissues and organs.
Jun 09		**DNA and the transmission of information** **The ways in which different organisms use inorganic ions**

APPENDIX 3: SUGGESTED ESSAY TITLES TO PREPARE

I've included a list of possible titles which I think reflect the changes of direction within the new A level. Each one is very synoptic and covers a wide range of topics from AS and A2.

1. Lipids in health and disease.
2. Genes and diversity.
3. The ways in which different organisms become adapted to their environments.
4. Coordination within organisms and between organisms and their environments.
5. Discuss how scientists collect, analyse and interpret biological data.
6. A space probe brought back samples of life-forms from a hot, dry planet with low atmospheric oxygen but high carbon dioxide concentrations. Describe the adaptations these life-forms would have in order to survive these conditions.
7. The physiological impact of lifestyle on health.
8. The impact of human activities on the diversity of animals and plants.
9. Receptors and their roles in coordination.
10. The pathways of synthesis of carbohydrates from atmospheric carbon dioxide.
11. Proteins such as insulin, FSH or an anti-influenza antibody are made in cells that are remote from their target tissues. Describe how one of these is synthesised and exerts an effect elsewhere.
12. Perform a critical analysis of methods used to collect biological data.
13. Stem cells research offers a great number of potential benefits to humans. It also comes with many down sides. Write a balanced account of the ways in which stem cells could and should be used to benefit humans.
14. Discuss the benefits and drawbacks of gene cloning technologies.

ESSAY 01: THE DIFFERENT WAYS IN WHICH ORGANISMS USE INORGANIC IONS

Inorganic ions are charged particles that do not contain carbon atoms bonded together. While organisms are mainly built from carbon-containing molecules, their functions rely on inorganic ions such as nitrate, hydrogen and calcium. This essay will detail some of the roles of specific ions and describe how animals, plants and bacteria use them.

Productivity in an ecosystem in the soil is limited in part by the availability of fixed nitrogen in the soil. Nitrogen fixing bacteria in the roots of leguminous plants reduce atmospheric nitrogen to ammonium using ATP and reduced NAD. The ammonium ions released into the soil are oxidised by nitrifying bacteria firstly to nitrite, and then to nitrate. This oxidation increases the nitrogen content in the soil which plants can use to produces many useful molecules including amino acids, proteins, DNA and ATP. The formation of these ions forms part of the ecological nitrogen cycle which plays a key role in sustaining life on this planet.

Plants are the producers for an ecosystem. They photosynthesise carbon dioxide and water and produce energy in the form of carbohydrates and other molecules. Photosynthesis requires water, and plants gain water from the soil using mineral ions such as nitrate produced by the nitrifying bacteria. Hydrolysis of ATP releases energy for processes such as active transport of the nitrate ions (and others such as potassium etc) from the soil into root hair cells, a process that lowers water potential and is used to draw water into the plant from the soil. In leaves, photosynthesis involves the photolysis of water, a process that involves the attachment of two electrons to a magnesium ion in chlorophyll and the production of hydrogen ions from the breakdown of water. Together with the electrons, the hydrogen ions are used to reduce NADP in the light-dependent reaction in the thylakoid. The hydrogen ions and electrons in turn are used to reduce glycerate-3-phosphate to form triose phosphate and glucose. Hydrogen ions also play a role in the production of ATP in the electron transport chains. They are pumped into the inter-membrane space and generate an electrochemical gradient that provides energy for the activation of ATPase which combines ADP and inorganic phosphate ions to form ATP.

The glucose, proteins and other molecules produced by the plants can then be consumed by animals for use in their life processes. The glucose undergoes respiration in cells in three different stages, each involving inorganic ions. On hydrolysis, ATP releases energy and a phosphate ion which can be used to phosphorylate glucose in the cell cytoplasm during its glycolysis. This phosphorylation makes the glucose more reactive and prevents it from leaving the cell. Following the transfer of hydrogen ions to coenzymes such as NAD, the pyruvate formed enters the mitochondrion and is decarboxylated and oxidised, in the process transferring its hydrogen ions and electrons to NAD and FAD. These hydrogen ions are pumped into the inter-membrane spaces of the cristae and are used to create an electrochemical gradient to form ATP as part of oxidative phosphorylation.

On role of The ATP produced is in the formation of a resting potential in nerve cells. Hydrolysis of ATP provides energy that is used to pump out three sodium ions and pump in two potassium ions into the axon of a neurone through a specific cation pump by active transport. A reduction of the membrane permeability to sodium ions maintains a resting potential of -70mV on the inside of the axon. Generation of an action potential also uses the charges from ions. Sodium gated channels open in the axon membrane allowing sodium ions to enter. This causes the membrane to depolarise until the threshold voltage of +40mV opens potassium gated channels. This causes potassium ions to leave repolarising, and eventually hyperpolarising the cell. This wave of depolarisation caused by these ion movements allows the passage of nerve impulse and coordination of the animal within its environment, allowing it to move sense and move effectively.

This movement involves the contraction of muscles, another process that uses ions, this time calcium. Calcium ions bind to tropnin, which causes tropomyosin to move away from the myosin head binding site on actin filaments. Once an actomyosin cross-bridge is formed and the actin filament slides into myosin, calcium ions activate ATPase to hydrolyse ATP to ADP and phosphate ions, a process that releases energy for the detachment and reformation of cross bridges. Contraction of the muscle sarcomere allows the contraction of skeletal muscle, allowing the animal to move. Muscles contractions are also used by animals in processes such as controlling light entry into the eye blood flow in arterioles in maintenance of homeostasis. All these processes require nervous coordination and contraction, emphasising the importance of the inorganic ions for proper function.

Contraction of intercostals muscles allows ventilation of the lungs to take place in mammals. This introduces oxygen to the gas exchange surface, the epithelium of the alveoli of the lungs. In order to maintain a high concentration gradient, the oxygen is rapidly removed, a process involving another mineral ion, iron. Iron 3+ ions are attached to haem groups on haemoglobin inside red blood cells. The iron can form bonds to oxygen, allowing haemoglobin to load oxygen in the lungs when the partial pressure of oxygen is high. Each molecule of haemoglobin can bind four oxygen molecules allowing a rapid saturation and the production of oxyhaemoglobin. On contraction of the ventricles, the pressure forces the red blood cells through the body to regions where the partial pressure of oxygen is lower. Here, the haemoglobin unloads, making oxygen available for aerobic respiration and the production of ATP.

In summary, inorganic ions are used in a diverse range of functions in living organisms. This essay has described some of these roles learned during the a-level study and has stressed their key importance in the life processes of bacteria, plants and animals.

ESSAY 02: DNA AND THE TRANSMISSION OF INFORMATION

Deoxyribonucleic acid, DNA, carries the genetic code for all living organisms on this planet. It is variation in the information it carries in form of genes and alleles that produces the wide diversity of life, and the variations within and between species. This essay will describe the structure of DNA and illustrate the ways in which the information encoded within it is transmitted within a cell, and between cells and organisms.

DNA is a polymer, a double helix of two polynucleotide strands bonded together by hydrogen bonds. Each nucleotide comprises of a phosphate group attached to a five carbon deoxyribose sugar and an organic base containing nitrogen. These bases can be either adenine (A), thymine (T), cytosine (C) or guanine (G). Adjacent nucleotides are joined by a condensation reaction to form the phosphate-sugar backbone of a polynucleotide strand. Two complementary strands then join by specific base pairing (A to T, C to G), which then wind together to form the double helix which provides strength and stability to the molecule.

The information in DNA is encoded in the sequence of bases along the template strand of the DNA. A gene is a sequence of bases on DNA that codes for the sequence of amino acids in a polypeptide chain. Since proteins determine the functions and structures of cells, it is the DNA code that controls all cellular activities. Organisms of the same species carry the same genes at fixed positions, called loci, but individuals carry different slightly different versions, termed alleles. Variation in these alleles results in intraspecific variation within a species, such as blood groups, eye colour etc.

In order for the genetic material to be transferred into daughter cells as the organism grows or repairs, the DNA must be replicated by semi-conservative replication. The enzyme, helicase binds to the DNA breaking the hydrogen bonds allowing the exposed bases on the two template strands to be revealed. DNA-nucleotides then bind to exposed bases by specific base pairing with hydrogen bonds. DNA polymerase then joins adjacent nucleotides with a condensation reaction forming the phosphate-sugar backbone. Each of the two new DNA molecules formed each contains one of the original strands of DNA. In this way the replication is semi-conservative and helps to minimise the incidence of mistakes, termed mutations, in the copying of the code.

In prokaryotic cells the DNA is free in the cell cytoplasm, but in eukaryotes it is bound within a nucleus and joined to structural proteins, called histones. The structure formed is called a chromosome and it is these that must be separated for the daughter cells to carry the same genetic information as the parent cell.

The process of mitosis separates the two copies of each chromosome. During prophase the chromosomes coil up and become visible, the nuclear envelope disappears and the chromosomes attach to spindle fibres at the equator of the cell using their centromere in metaphase. In anaphase the centromere divides and the spindle contracts drawing the chromatids to opposite poles of the cell. After telophase and cytokineis, two new daughters are formed, each containing an identical copy of the DNA code; hence the encoded information has been transmitted vertically. Bacteria also possess the ability to transmit some of their genes horizontally. Conjugation tubes can form between two bacterial cells and the plasmid, small loops of DNA that carry codes for antibiotic resistance, can pass between the two cells. So if one bacteria owns has a plasmid that carries the code for penicilinase, the plasmid can be replicated and passed via conjugation to another. Now both cells are resistant to penicillin.

The information on DNA is encoded as triplets of bases, called codons. Each triplet can code for one amino acid in a polypeptide chain. So for example, if GCA codes for the amino acid alanine and TAC codes for glycine, then the code GCAGCATACGCA would code for a polypeptide with the sequence ala-ala-gly-ala. As there are over twenty different amino acids in nature, a triplet code allows coding of up to 64 amino acids. Such a code is termed redundant and in reality each amino acid is coded for by several different codes. This minimises mutation rates as, for example, if GCC also codes for alanine, then a mutation from CA to CC would have no effect on primary structure. Each of the codons is translated in sequence as the code is non-overlapping, but first the genetic information must be transcribed, and then transferred out of the nucleus. It is transferred as an RNA molecule, a single-stranded polynucleotide containing the base uracil instead of thymine, and the five-carbon sugar ribose.

Transcription produces a copy of a gene in the form of messenger RNA (mRNA). Helicase binds to the gene locus causing DNA to unwind and reveal a template strand. RNA-nucleotides bind by specific base pairing and RNA polymerase joins them by condensation to form a strand of pre-mRNA. Introns (non-coding regions) are then removed and the exons (coding regions) are spliced together with enzymes to form the mRNA which is small enough to diffuse through the nuclear pore and bind to a ribosome on the rough endoplasmic reticulum.

The process of protein synthesis, or translation can now begin. In the cytoplasm, a transfer RNA (tRNA) molecule binds to a specific amino acid and two such complexes deliver their amino acids to the ribosome. The anticodon on tRNA binds to the complementary codon on mRNA by specific base pairing (A to U, C to G). An enzyme now forms the peptide bind between the amino acids by condensation using energy from ATP and the process is repeated building up the polypeptide chain. Alterations of the base sequence of the gene, mutations alter the structure of the mRNA and so possibly altering the primary structure of the polypeptide coded for. These can be substitutions, deletions of additions. The greatest corruption of the code occurs with the latter two which cause frame shifts that are catastrophic to the base sequence and the primary structure of the coded protein.

Sexually reproducing organisms transmit their genes in the form of haploid gametes (ova and sperm, or pollen) formed by meiosis. This reductive cell division halves the chromosome number so the diploid number of chromosomes can be regenerated on fertilisation. Meiosis introduces variation through crossing over and independent segregation of chromosomes, and random fusion ensures further variety in the offspring produced. In this way the genetic information is transmitted from generation to generation introducing a diverse range of alleles that adds not only variety, but helps ensure a population can survive and adapt to any environmental changes.

ESSAY 03: THE PART PLAYED BY ENZYMES IN THE FUNCTIONING OF DIFFERENT CELLS, TISSUES AND ORGANS

Enzymes are biological catalysts that control almost all chemical reactions inside and outside cells. In this way they control the functions of not only individual cells, but of collections of cells (tissues), or collections of tissues (organs). This essay will demonstrate the diverse range of ways enzymes contribute to the functioning of these structures.

Enzymes are globular proteins which have a specific tertiary structure that has a complementary shape to that of a specific substrate molecule. The lock and key model is used to describe enzyme action. For example the enzyme lactase has an active site (a lock) that is complementary only to lactose (the key). Sucrose, a similar disaccharide has a different shape to lactose and so cannot bind to lactase's active site. On binding to the active site, an enzyme-substrate complex is formed and reaction takes place. The products have a different shape and can no longer remain bound. In the induced fit model, the active site is not complementary to the substrate, but on binding the shape changes and the active site forms, molding itself to the substrate a tight glove would mould to a hand.

Humans gain the molecular building blocks they need for energy and growth from digestion of food by the digestive system. This is a system of organs that is adapted for the hydrolysis of food molecules and the absorption of their products. In the mouth, the enzyme amylase in the saliva hydrolyses starch to the disaccharide maltose, which is further digested in the intestinal epithelium to α-glucose. In the stomach endopeptidases such as pepsin break down proteins in smaller peptides, and exopeptidases such as trypsin further hydrolyse these into amino acids in the small intestine. Glucose is then absorbed by sodium-glucose transport, a type of active transport that involves the enzyme ATPase which hydrolyses ATP to ADP and Pi releasing energy to pump sodium ions out, and potassium into epithelial cells creating diffusion gradient for sodium and glucose uptake. Enzymes also play a key role in digestion of large insoluble food molecules into smaller, more soluble products that can be transported and assimilated in fungi and bacteria. Decomposers in the ecosystem, the fungi and bacteria, release hydrolytic enzymes such as lipase, carbohydrase and protease (to digest triglycerides, carbohydrates and proteins respectively). The soluble products of this extracellular digestion (e.g. fatty acids, glucose, and amino acids) can then be absorbed and assimilated into useful compounds.

All organisms carry the genetic code for their functions as a DNA molecule. Before a cell divides by mitosis, the DNA must undergo semi-conservative replication to produce two identical copies for the daughter cells. Enzymes play a key role here. Helicase binds to the DNA, breaking the hydrogen bonds that hold the two polynucleotide chains together. This reveals two template strands which have exposed bases which bind to DNA-nucleotides. A second enzyme, DNA polymerase then forms a phosphate-sugar backbone by joining adjacent nucleotides with a condensation reaction.

Some cells, such as β-cells of the pancreas synthesise and secrete protein hormones such as insulin. In order for the genetic code on DNA to be expressed and the insulin formed, the DNA must be transcribed as a pre-mRNA molecule, spliced to form mRNA and transcribed as a protein. Enzymes are involved in each step. Helicase binds to the gene locus and cause the gene to unwind exposing the template strand. RNA polymerase joins adjacent nucleotides in a condensation reaction to form the pre-mRNA strand. Enzymes in the nucleus remove non-coding introns, and splice together the coding exons leading to the formation of an active mRNA which binds to a ribosome on the rough endoplasmic reticulum. Transfer RNA complexes line up with their anticodons on the codons on mRNA and bring two amino acids in contact with an enzyme in the ribosome that condenses them together by forming a peptide bond. The process is repeated to build up the primary structure of the insulin molecule. The action of the hormone insulin also involves phosphorylase enzymes which cause the condensation of glucose molecules into the storage polysaccharide glycogen in the liver by glycogenesis.

All living cells release the energy in substrate molecules using aerobic or anaerobic respiration. The respiratory process is a sequence of interconnected enzyme controlled steps called a metabolic pathway. Other pathways include photosynthesis and the synthesis of steroid hormones such as oestrogen from cholesterol. During glycolysis, the link reaction and the Krebs cycle, some of the steps include oxidation by dehydrogenase enzymes. This oxidation involves the transfer of hydrogen ions and electrons from the substrate and passing them to a coenzyme which becomes reduced. For example, in the cytoplasm, when triose phosphate molecules are oxidised to pyruvate as part of glycolysis, the coenzyme NAD is reduced forming reduced NAD. The coenzyme forms part of the active site of the dehydrogenase enzyme allowing it to function as a catalyst and be reformed.

The ATP formed as part of respiration is used in a wide variety of contexts in biology. For example in order for an animal to move and hunt for food within its environment, it has to contract its muscle tissue. The tissue is composed of cells containing actin and myosin filaments which move relative to each other to contract a sarcomere. For this to happen, actomyosin cross-bridges form between the actin and myosin. Once activated by calcium ions, the enzyme ATPase then hydrolyses ATP to ADP and Pi releasing energy for the detachment and formation of more cross-bridges, giving rise to the sliding filament theory of muscle contraction. This enzyme also helps release energy from ATP in a wide variety of contexts, such as in the active transport of sodium ions out of an axon through sodium-potassium cation pump in the generation of a resting potential, or in the active transport of nitrate ions into a root hair cell to lower water potential to draw in water to generate a root pressure.

This essay has established that enzymes are fundamental biological molecules which offer a diverse range of functions to living organisms.